Paulo Roberto de Freitas da Silva

Archimedes' Principle

Paulo Roberto de Freitas da Silva

Archimedes' Principle

A Proposal for Secondary Schools

ScienciaScripts

Imprint

Cover image: www.ingimage.com

This book is a translation from the original published under ISBN 978-3-330-75702-8.

Publisher:
Sciencia Scripts
is a trademark of
Dodo Books Indian Ocean Ltd. and OmniScriptum S.R.L publishing group

120 High Road, East Finchley, London, N2 9ED, United Kingdom
Str. Armeneasca 28/1, office 1, Chisinau MD-2012, Republic of Moldova, Europe
Printed at: see last page
ISBN: 978-620-8-28500-5

SUMMARY

DEDICATORY

To God, creator of everything, always present, who guides me and supports me in the most difficult moments.

To my family, especially my mother, Valdenora.

To Tatiana Rodrigues, for her eternal presence in my life, companionship, emotional and intellectual support and for always believing in my potential.

To Antongnioni Melo, for her invaluable friendship, companionship and presence in my life.

To all those who have made me grow intellectually and as a human being.

ACKNOWLEDGMENTS

First of all, I thank God, creator of heaven and earth and of all things visible and invisible, for giving me the gift of life and for allowing me to remain firm and convinced of his existence.

To my dear mother, Valdenora, for the support she has always given this son of hers. Educating, guiding, loving... My greatest and best example of virtue. My inspiration... My most valuable treasure.

To Antongnioni Melo, my best and noblest friend, who always gave me attention, support, encouragement and patience. The best advisor I could ever have...

To Tatiana Rodrigues, for the priceless hours she spent by my side, supporting, advising, guiding, believing in my academic potential? Being my north. My compass, in stormy times and also in the calm.

Introduction

A major problem observed in education and involving the teaching-learning process of high school physics is the model used to teach this subject. Currently, a considerable number of secondary schools use the traditional chalk and board method as their only teaching methodology. In addition, there is an excessive use of formulas and mathematical equations, thus valuing passive learning, and there is little use of practical experiments that relate Physics to the students' daily lives or other playful strategies. This ends up accentuating the lack of interest on the part of the students, which is already visible in secondary school classes among a considerable proportion of this group. As Ricardo notes (*apud* CLEMENT; CUSTÓDIO; ALVES FILHO, 2015. p. 102): "Taking the teaching of Science, in general, and Physics in particular, as an example, there is significant complaint and denunciation on the part of teachers, of a lack of interest and motivation on the part of students to study and learn Physics".

On the other hand, an experimental approach that brings everyday applications into the classroom could optimize the learning of physical concepts and also allow students to adopt a more active stance, thus making them the protagonists in the construction of knowledge, which would be more cognitively interesting.

With regard to the phenomenon to be addressed in this work, it is known that around two thousand years ago, Archimedes (287-212 BC) was

was tasked with solving a problem of great magnitude. The problem was posed by Herasius, who suspected that his goldsmith, when making his crown, had deceived him by saying that it was made of pure gold. As Heraclius believed that the crown was not made of pure gold, the ruler proposed that Archimedes prove or disprove his suspicions. Archimedes, who was a famous inventor, physicist and mathematician in Syracuse, a city in Italy, spent several days thinking about how to prove that the crown was made entirely of gold or not.

Until one day, when Archimedes was bathing in a bathtub, he realized that there was a relationship between the amount of water his body drained and the fact that his body felt lighter when he was in the water.

In fact, this relationship does exist. A solid that is immersed in a liquid sinks if it is denser than the liquid, and its measured weight when immersed is equal to its real weight minus the weight of the liquid displaced by it. This is the statement of Archimedes' principle. The force present when the liquid is immersed, which decreases its weight, is called the buoyant force.

A major problem in the physics teaching and learning process is the current model used to teach this subject. Currently, although this problem is not new, with rare exceptions, most secondary schools still use the traditional teaching method: blackboard and chalk, an excess of formulas and mathematical equations, and a minimum of experiments that relate Physics to the students' daily lives. The use of this single methodology is often insufficient for proper learning, since it does not take into account the individual differences, needs and difficulties of the students, thus making its reach very limited. This was analyzed by Laburù, Arruda and Nardi (2003):

> As is widely known in the field of science education, the old teaching strategies of blackboard and chalk, linked to the old coercive and exclusive objectivist pedagogical paradigm, based on the logic of the "giving" of knowledge, which privileges listening over speaking, are insufficient to ensure that learners really learn scientific concepts (LABURÙ; ARRUDA; NARDI, 2003, p. 248).

Faced with this reality, efforts have always been made to change the situation. Evidence of this can be seen in the wide-ranging discussions that have taken place in academic circles and in publications in the field, regarding the best way to approach the content of Physics in secondary schools. Currently, a good number of Brazilian universities have their own study groups that discuss this issue, with a notable example being the Physics Teaching Rework Group (GREF) at the University of São Paulo (USP), as well as groups of researchers involved in lato sensu and stricto sensu postgraduate programs. Outside Brazil, we can mention programs such as the Physical Science Study Committee (PSSC), a North American project developed in the 50s, the Project Physics Course (popularly known as the "Harvard Project"), developed at Harvard University in the 60s, with ideas similar to the PSSC, and the Conceptual Physics project, by North American physicist Paul G. Hewitt, which deals with physics through concepts, avoiding the use of formulas, created at the end of the 90s. As Araujo and Abib (2003, p. 176) point out:

> The difficulties and problems affecting the education system in general and particularly the teaching of physics are not new and have been diagnosed for many years, leading different groups of scholars and researchers to reflect on their causes and consequences.

These few examples illustrate the efforts, not only local but also global, to improve and make the teaching of physics not only more enjoyable, but also more cognitively meaningful and less traumatic for secondary students, as well as the dimension that this subject reaches in academic circles.

Given this situation, it can be seen that the use of the traditional methodology based on the use of the blackboard and chalk as the only teaching-learning strategy seems to be insufficient for complete and effective learning on the part of secondary school students. As such, there is an urgent need to adopt methodologies that can provide greater support to the traditional teaching method. It therefore seems natural to adopt a pluralist methodology for science education, as taught by Laburù, Arruda and Nardi (2003), because a pluralist pedagogical proposal for science education starts from the assumption that every teaching-learning process is highly complex, changes over time and involves multiple types of knowledge and is far from trivial.

Still on the subject of methodological pluralism, we can cite the thoughts of the notable Austrian philosopher Paul Feyerabend. He advocated methodological pluralism in scientific methods and was one of the great critics of the use of a single scientific methodology in academic circles. According to Feyerabend (2011):

My intention is not to replace one set of general rules with another set of the same kind: rather, my intention is to convince the reader that *all methodologies, even the most obvious ones, have their limits.* (FEYERABEND, 2011, p. 47, emphasis added).

Expanding on Feyerabend's ideas at this point, could a methodological pluralism not be used in the teaching of Physics? Couldn't the use of more than one methodology, something that goes beyond the "blackboard and chalk", make learning Physics more attractive and meaningful for the student?

In view of the reality found and already exposed in high school classes today, we propose to provide students in this segment with a pluralistic approach to the subject of Hydrostatics - Archimedes' Principle. The main focus of this approach is experimentalism, when the qualitative aspects of Archimedes' Principle are emphasized. In conjunction with the proposed experiment, a more theoretical approach could also be envisaged, where both qualitative and quantitative aspects would be dealt with, all complemented by the historical aspects that permeate this principle. The choice of an experimental activity as a strategy for teaching Physics, with the aim of assisting traditional teaching (blackboard and chalk), stems from the fact that the use of experiments has been pointed out by teachers and students as one of the most fruitful ways of minimizing the difficulties encountered in learning and teaching Physics in a meaningful and consistent way (MORAES and MORAES, *apud* ARAÙJO and ABIB, 2003, p. 176).

Once it was decided to carry out an experiment that focused on the study of Archimedes' Principle, we tried to prioritize an apparatus that was simple to build and execute, as well as being low-cost, since it takes into account that the resources found in public schools are often inadequate, scarce, precarious or even non-existent.

The main aim of this work is not to replace the traditional lecture. On the contrary, the aim is to enrich the students' education process by providing an additional approach, in this case the application of a simple experiment, with the intention of helping students learn Archimedes' Principle and other concepts that relate directly and indirectly to this principle. The aim is also to record the different stages involved in making the experiment, starting with the material to be used to make it and moving on to its actual manufacture.

The experiment is a priori aimed at secondary school students, but with a few adaptations, it can be used at other stages of education and with students of varying levels of knowledge and different age groups.

At the end of the experiment, it is hoped that the students will be able to grasp the physical concepts covered in a more solid and complete way, with emphasis on the concepts of force, thrust, mass, density, pressure and more general qualitative characteristics of fluids. It is also hoped that the students will see experimental treatment as an important tool for describing phenomena in the sciences in general. The aim is also to promote a discussion in the school environment about the benefits of low-cost teaching laboratories, where it would be possible for the school community itself to produce simple experiments that are easy to carry out, without the need for expensive, large and complex experimental apparatus, in order to maximize students' cognitive development, using more playful and attractive strategies.

The experiment to be used in this project focuses on making a small submarine out of low-cost materials, easily found in stores, often available even in the student's own home, and which can be built and reproduced easily by most students. It is hoped that after carrying out the experiment, the students will become more active subjects, taking the lead in the construction of their knowledge and not just listeners and mere supporting actors.

Chapter 1

Archimedes' principle and the science laboratory

There are countless legends surrounding the life story of Archimedes, the remarkable Greek physicist, mathematician, philosopher and inventor, who lived in Sicily between 287 BC and 212 BC. However, when studying the facts surrounding Archimedes' life, the best-known story is that of the problem posed to him by Heranus, king of Syracuse, where this important scientist lived. According to Nussenzveig (2002):

> According to the legend told by the historian Vitruvius, Heranus, king of Syracuse, suspected that he had been tricked by a goldsmith, who had mixed silver in the making of a gold crown, and asked Archimedes to check it out: "While Archimedes was thinking about the problem, he happened to arrive at the public bath, and there, sitting in the tub, he noticed that the amount of water that overflowed was equal to the immersed portion of his body. This suggested to him a method of solving the problem, and without delay he leapt joyfully out of the bath and, running home naked, shouted loudly that he had found what he was looking for. For, as he ran, he repeatedly shouted in Greek 'eureka, eureka' ('I have found it, I have found it')." According to the historian, by measuring the volumes of water displaced by gold and silver and by the crown, Archimedes was able to prove the forgery (NUSSENZVEIG, 2002, p. 11).

However, Archimedes' genius goes far beyond the famous and brief episode recounted above. He was a crucial figure in the defense of Syracuse against the Romans and was responsible for countless contributions in various fields of knowledge. According to Barbosa and Breitschaft (2006):

> [...] throughout his life, he contributed to various areas of human knowledge (mathematics, physics, engineering, astronomy and philosophy). Archimedes was famous in his time for inventing various machines used in the defense of Syracuse against the invasion of the Romans, during the first two Punic Wars. He is considered by some historians of mathematics to be one of the greatest mathematicians in history, along with Newton, Euler and Gauss. In physics, he is credited, among others, with the following contributions: the Law of the Lever, in relation to which he is said to have commented *Give me where to stand and I'll move* the *Earth*; the beginning of *pycnometry* (from the Greek *pyknós*,

meaning dense, compact, thick), which is a methodology for measuring the volume or specific mass of solids or liquids; and the Law of Thrust or Archimedes' principle. [...] (BARBOSA and BREITSCHAFT, 2006, p. 115, emphasis added).

These brief accounts show us the beginnings of what is now known as Archimedes' Principle, which is of unique importance to the sciences in general and has countless practical and everyday applications.

1.1 Thrust

When trying to lift an object that is completely submerged out of the water, this task becomes much easier than if it were done out of the water. This can be explained by the concept of thrust, which can be thought of as an apparent loss of weight when objects are immersed in a fluid medium. For example, lifting a large piece of rock from the bottom of a lake is relatively easy as long as the rock is below the surface, i.e. submerged. However, when this same rock is removed from the surface, the force to continue lifting it increases considerably, making the task impossible depending on the size of the object.

This situation can be explained by the fact that when an object is submerged in a liquid, the liquid exerts an upward force on the object, which is the opposite of the gravitational force of attraction. This upward force is called buoyancy and is a consequence of the increase in pressure with depth.

To determine the characteristics of thrust, we will consider a cylindrical body of volume V_c, base area A and height h, immersed in a fluid that is in equilibrium with density ρ, in a place where the acceleration of gravity is equal to .

It can be seen that at any point on the surface of the cylinder, as a result of the pressure that the fluid exerts on the object, forces appear that are perpendicular to its surface. These horizontal forces (manifesting themselves in pairs) acting on the surface of this solid are at the same depth and are therefore produced by columns of fluid of the same height, which is why they have the same intensity, thus balancing each other two by two. We can therefore conclude that the value of the thrust on the body will be equal to the difference between the intensities F_B and F_A of the forces exerted by the fluid on the bases of the cylinder under study, as these do not cancel each other out. It is important to note that the forces at the top have lower intensities than at the bottom. The resultant of these forces tends to push the cylinder upwards. This resultant force is what is known as thrust.

The situation described is shown below:

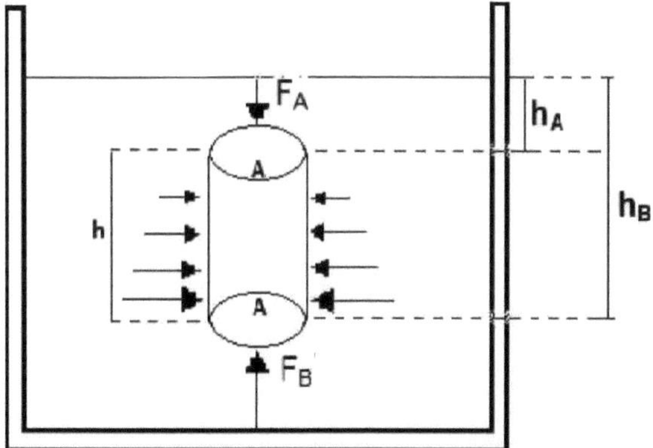

Figure 1 - Archimedes' principle Source: author's own work

This means that

$$E = F_B - F_A$$

Assuming that the atmospheric air column, i.e. the area above the fluid column, is subject to a pressure p_0 , the pressures at the bases A and B of the cylinder under analysis will be equal to:

$$p_A = p_0 + \rho g h_A \quad e \quad p_B = p_0 + \rho g h_B$$

It is known that:

$$p = \frac{F}{A} \quad \Rightarrow \quad F = pA$$

$$F_A = p_A A \quad e \quad F_B = p_B A$$

So there you have it:

$$E = F_B - F_A$$

$$E = p_B A - p_A A$$

$$E = A(p_B - p_A)$$

$$E = Ap_B - Ap_A$$

$$E = A(p_0 + \rho g h_B) - A(p_0 + \rho g h_A)$$

$$E = Ap_0 + A\rho g h_B - Ap_0 - A\rho g h_A$$

$$E = A\rho g h_B - A\rho g h_A$$

$$E = A\rho g(h_B - h_A)$$

Since, $h_B - h_A = h,$, it follows that:

$$E = \rho g \cdot hA \quad , \quad onde\ hA = V_c$$

$$E = \rho g V_c$$

As the volume V_c of the cylinder analyzed is equal to the volume of fluid displaced by it, the following relationship is obtained:

$$E = \rho g V$$

Where:

E: thrust force

ρ: fluid density

g: acceleration due to gravity

V: volume of fluid displaced

Since the product of the density (ρ) of the fluid by the volume of the fluid (V) displaced is the mass m of the fluid displaced, that is:

$$\rho = \frac{m}{V} \quad \Rightarrow \quad m = \rho V$$

It follows that:

$$E = mg \quad \Rightarrow \quad E = P$$

This is the general statement of Archimedes' Principle, quoted by Doca, Biscuola and Bôas (2010):

> When a body is totally or partially immersed in a fluid in equilibrium under the action of gravity, it receives a force from the fluid called **thrust** (or Archimedes' impulse). This force always has a vertical direction, from bottom to top and an intensity equal to the weight of the fluid displaced by the body. (DOCA; BISCUOLA; BÔAS; 2010, p. 401, emphasis added).

This formalizes Archimedes' Principle, which basically deals with the response that a fluid gives to the presence of a body when immersed in it.

1.2 Floating

As discussed earlier, the resultant force acting on a body submerged in a fluid is the sum of the body's thrust and weight. Let's suppose that an object is inserted into a fluid so that it is completely submerged. If the object is then abandoned, the forces acting on it will only be its own weight (P) and the buoyancy (E) exerted by the liquid. In this situation, three different cases can be analyzed:

1. **The thrust will be less than the weight of the object** ($E < P$). In this first situation, the resultant of these forces will be directed downwards and the object will sink until it reaches the bottom of the container. This is the situation that occurs if a rock is thrown into the water, for example.

2. **The value of the thrust is equal to the weight of the object** *(E = P)*. In this second situation, the resultant of the forces will be zero and the object will be at rest in the position in which it was abandoned. This situation is what happens to a submerged submarine at rest at a certain depth.

3. **The value of the thrust will be greater than the weight of the object** *(E > P)*. In the last situation analyzed, the resultant of the forces will be directed upwards and the object analyzed will rise through the fluid.

The three situations studied above are illustrated in Fig. 2 below:

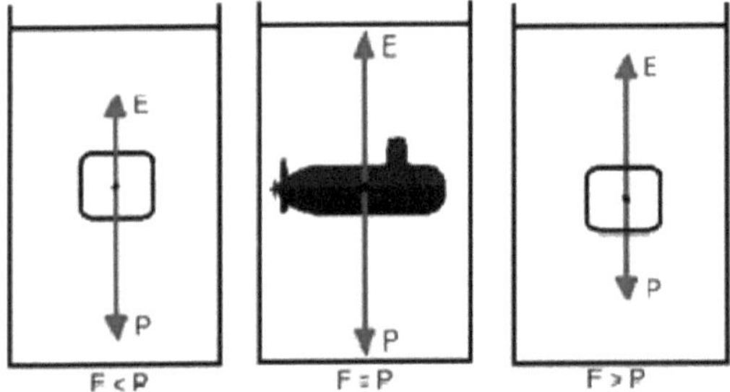

Figure 2 - Floating conditions of an object in a liquid medium

Source: author

It can also be seen that as long as the object is completely submerged, *E > P*. When the object begins to reach the surface of the liquid, the amount of liquid displaced by the object begins to decrease and, consequently, the value of the thrust (E) also decreases. At a certain point, the object will be displacing a quantity of liquid whose weight (*P*) will be equal to its own weight, i.e. *E = P*. Once this situation is reached, the object will float, in equilibrium, because the resultant of the forces acting on it is zero. The situation described is illustrated in Fig. 3, shown below:

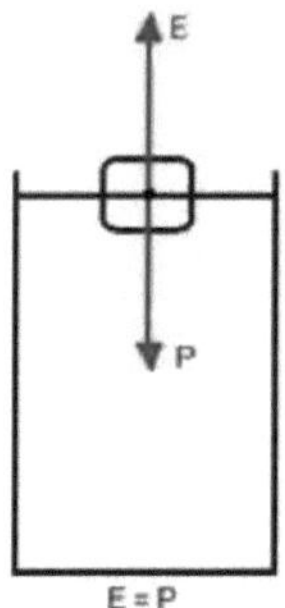

Figure 3 - Floating conditions of an object in a liquid medium (E = P) Source: author's own work

Still on the subject of the buoyancy of bodies, at this point, if we analyze how this buoyancy is influenced by the density of the fluid in which the body is immersed, we know that according to Archimedes' principle, the buoyancy is equal to the weight of the liquid displaced. Thus:

$$E = P$$

$$E = m_d g$$

Where ρ_l is the density of the liquid and V_d is the volume of the liquid displaced, we have:

$$\rho_l = \frac{m_d}{V_d}$$

$$m_d = \rho_l V_d \quad \therefore \quad E = \rho_l V_d g$$

It can be seen from this expression that the value of the thrust will be greater the greater the volume of liquid displaced and the greater the density of this liquid.

With regard to the weight (P) of the body submerged in the liquid, it can be expressed as a function of its density (ρ_c) and its volume (V_c):

$$P = mg$$

$$\rho = \frac{m}{V}$$

$$m = \rho V$$

$$m = \rho_c V_c$$

$$P = \rho_c V_c g$$

Thus, when a body is totally submerged in a fluid, it will be displacing a volume of liquid V_d equal to its own volume V_c , i.e. $V_d = V_c$. Therefore, for a body totally immersed in a fluid, the two relationships can be compared:

$$E = \rho_l V_c g$$

$$P = \rho_c V_c g$$

Analyzing the two ratios, we can see that both differ only in the values of ρ_l (density of the liquid) and ρ_c (density of the body).

Thus, returning to analyze the three conditions for floating an object in a fluid medium, we have that:

1. **If $\rho_l < \rho_c$, then $E < P$.** Under these conditions, as **has** already been observed, the body studied will sink into the liquid.

2. **If $\rho =_l \rho_c$, then $E = P$.** Under these conditions, as has already been analyzed, the body will be in equilibrium when it is completely submerged in the liquid.

3. **If $\rho_l > \rho_c$, then we have $E > P$.** This condition, as **we have** already analyzed, in which the body will rise in the liquid, until it reaches an equilibrium position on its surface, being partially submerged, where $E = P$.

With the analysis carried out, we can propose the study of two everyday situations: the submarine and the balloon. In the case of a submarine, if this vehicle is submerged, in equilibrium, as shown in Fig. 2, we can conclude that its density is, on average, equal to the density of the fluid in which it is immersed, i.e. water. With similar reasoning, we can conclude that a balloon rises into the atmosphere because its average density is lower than that of atmospheric air, a situation illustrated in Fig. 4 below:

Figure 4 - A balloon rises in the atmosphere due to buoyancy (E > P)

Source: author

With the extensively discussed analyses of the concept of buoyancy and the conditions under which bodies float when immersed in a fluid, it is possible to predict when a solid body will float or sink.

1.3 The science teaching laboratory

With regard to the importance of experimental practice in general, its value is unquestionably recognized. Regardless of whether it is carried out in school environments or not, it is known that the use of experiments as an instrument for questioning existing and already conceived truths, for reflecting on already established ideas and for reformulating existing concepts by models that are better adjusted to reality are widely known throughout time, as the history of humanity itself shows us. This fact is pertinently observed by Braga, Guerra and Reis (2011):

The experimental research that was developing in the first half of the 13th century increasingly posed questions to natural philosophers that challenged established truths.

When investigating the temperature of a thermometer immersed in a container of alcohol, for example, it was found that when a portion of the alcohol evaporated, the temperature recorded by the thermometer decreased, indicating that the liquid had become colder in the process. As the effect occurred even in a vacuum, the then current theory of evaporation - which admitted that the phenomenon was due to the liquid dissolving in air - needed to be reformulated. (BRAGA; GUERRA; REIS; 2011, p. 64)

The situation described demonstrates, in a way, the efficiency of a science built on the pillars of experimentation, as it allows for the revision and reformulation of previously conceived ideas and concepts by others that are more coherent, modern and based on more solid principles and thoughts than the previous ones.

From this point of view, if we transpose these ideas to physics and science teaching laboratories in general, wouldn't they also have the capacity to develop similar reflections and skills in the students involved in experimental practices? Couldn't such experimental practices, if experienced by students, undo misconceptions and ideas, improving them and replacing them with scientifically correct ones?

It is well known that experimental practice has a very different objective to that found in theoretical subjects. Instead of just exposing content and expecting students to grasp it in the best possible way, a methodology which makes the teaching-learning process less active on the part of the students, since most of the time the student becomes an assistant in the learning process, laboratories, on the contrary, foster and stimulate much richer learning and in a much more active, attractive and playful way, since the target audience lives and interacts directly with the proposed problem situations. As Ferreira (1993) shows us:

Bringing back the value of the didactic laboratory, i.e. experimental activities in general, could contribute to a better education for students, at least with regard to the phenomena of classical physics, and make them better able to deal with the explosion of knowledge in which they are immersed. (FERREIRA; 1993, p. 97).

This is because such environments and the experiments carried out in them, if planned and executed in a serious and committed manner by the players involved, give students the opportunity to learn, revise and make more solid the fundamental concepts, which before experimentation could become somewhat vague, abstract and obscure, thus paving the way for learning more complex concepts and ideas.

However, physics (and science) laboratories have an even more important purpose in terms of the teaching-learning process and cognitive development. They awaken and develop the ability to critically analyze problem situations and their results. They also allow for a much more active discussion of the meaning and validity of the situations proposed in the analysis, enabling students to draw their own conclusions, which are better grounded in science. On a positive note, it also encourages group and cooperative activity. Grandini and Grandini (2004) also teach us a lot about the aims of teaching laboratories:

The didactic laboratory allows students to experience and handle instruments that allow them to learn about different types of activities, and can stimulate their curiosity and desire to learn and experience science. The laboratory should encourage students to know, understand and learn to apply theory to practice, mastering tools and techniques that can be used in scientific research. They should learn to observe scientifically, interpret and analyze experiments through objectivity, precision, confidence, perseverance, satisfaction and responsibility (GRANDINI and GRANDINI, 2004, p. 252).

With regard to the low-cost physics teaching laboratory, which is the subject of this paper, the aim is to bring basic physics experiments into the classroom environment. These environments are mainly aimed at students in public schools, institutions which often have no laboratories at all, and when they do, they are often disused or in poor condition.

However, this argument does not limit the use of low-cost experiments only to the realities found in the country's public schools. Such experiments are also valid in schools with experimental resources, and are just as effective in these environments as they are in the reality of Brazilian public education, and can complement and enrich the apparatus already available, a fact discussed by Ferreira (1993, p. 100):

> [...] the use of low-cost materials is often seen as a solution for children or Third World countries. In reality, these statements are true, but they do not exhaust the possibilities and versatility that can be obtained with the use of such materials. At IF/USP we have been working for many years with low-cost materials in the subject of Instrumentation for Teaching Physics, in continuing education courses for teachers and with the Institute's students.

Ferreira (1993, p. 100) goes on to say that:

> [...] we can say that low-cost materials are not just for developing countries, but are used by all those who believe that they offer the possibility of seeing the phenomena in question more clearly. Shrigley (1971, p.361), comparing simple instruments with those found on the market, states that there are two levels to be taken into account: one refers to the efficiency of the device as a measuring instrument and the other is its didactic component. He says that it is important for students to know that instruments work "because there is a physical principle involved, and not because of the chrome they come out of the factory with".

In the light of the above, it stands out how positive the use of low-cost experiments can be in the reality of public education in the country, since this alternative opens up paths and possibilities in the field of experimentation in school institutions that are partially or totally devoid of such resources. Such experiments can thus support and complement theoretical subjects, helping them to learn concepts and ideas correctly, which before the experimental activity could have been misleading, taking the teaching and learning of physics in secondary schools to another level.

Chapter 2

Didactic Experiment Description

As has already been amply explained, the purpose of this work is to carry out a simple, low-cost experiment on the subject of hydrostatics, more specifically Archimedes' principle, as well as related concepts and ideas. As this is a low-cost didactic experiment, it is hoped that it can be applied with relative ease in a wide variety of educational institutions, including those without laboratories and other experimental resources, or where they are present, these are inoperative or in a poor state of repair.

In this section of the paper, we will present the experiment that aims to study Archimedes' Principle and some other concepts directly and indirectly involved with it. The proposed experimental setup can be seen in more detail in the sequence of photographs shown in Fig. 5 below:

Figure 5 - The experimental apparatus Source: author's own work

The material used, as well as the detailed script for building the apparatus illustrated in Fig. 5, can be found systematized in Appendices A and B, respectively.

The experimental apparatus consists of building a small submarine simulator using the materials listed in Appendix A. By means of this simple experiment, it is hoped that students will be able to verify the fact that the thrust of a given fluid medium (in the case of the experiment, water) acts from the bottom up on a body (in the case of the experiment analyzed, the submarine) that is immersed (submerged) in it. It also allows students to analyze the influence of the density of different fluids on the behavior of the submarine being analyzed.

By immersing the device under study in the container used with water, the apparatus must be filled with this fluid so that it becomes submerged. In general, the water gradually and naturally begins to invade the inside of the device, without the need for external intervention. It is important to keep the other end of the hose (tube) out of the water container, i.e. the one that is not inserted inside the submarine.

Once submerged in water and filled with this fluid, the device is expected to come to rest at some well-defined depth, as shown in Fig. 2. Of the three situations described in Theoretical Basis - Buoyancy, the one that can be observed at this point is situation number 2, in which case the submarine remains submerged at a certain depth. This means that

$$E = P$$

Or else:

$\rho =_{l} \rho_c$, *on average.*

Inflating air at the end of the hose facing away from the container of water, it is observed that as the air is introduced into the balloon and it inflates, the submarine under study expels the water with which it was completely filled, replacing this fluid with the air introduced into the balloon and it returns to the surface. In this case, situation number 3 is the one that best fits the analysis, which is also mentioned in the Theoretical Basis - Buoyancy. This means that in this situation:

$$E > P$$

This causes the submarine to return to the surface, and it can be said that:

$\rho_l > \rho_c$, *on average.*

The various situations discussed above deal with a qualitative analysis of the problem. Optionally, a more quantitative study could also be made possible, in order to broaden the didactic possibilities of the experiment, by proposing that students take measurements of pressure values as a function of height variation. This didactic situation could be made possible by grading some approximate values for the height of the fluid column (by marking the water container with masking tape, for example, and measuring with a ruler). Obviously, these calculations will be approximate, but it is believed that their didactic value will not be lost.

Another possible variation of this experiment, also allowing for a qualitative study, which is not discussed in detail in this paper, but which could make the analysis richer and more complete, consists of carrying out the same experiment, but inserting small "weights" inside the submarine. This would increase the average density of the

device (submarine), which would deal with another case already discussed: the situation in which $\rho_l < \rho_c$, i.e. a situation in which $E < P$. As already mentioned, this specific case would result in the complete submergence of the submarine, since in the resultant of the forces, the weight *(P)* would have a higher value than the thrust *(E)*.

The cases cited above illustrate suggestions of how the problem and the experiment can be approached. Of course, other approaches and studies can be used, either related to the topic or correlated with it, depending on the objective you want to achieve, since the experiment can be approached in different ways.

As far as the classroom practice is concerned, it took place in 4 lessons over two weeks (2 lessons per week of 50 minutes each). In the first two lessons, i.e. the first week, a theoretical approach was given to basic aspects related to Hydrostatics and Archimedes' Principle, such as the concepts of volume, density of a fluid, pressure and the concept of buoyancy, which was already presented as a force. Historical aspects related to the formulation of this principle were also dealt with in order to stimulate students' curiosity about the subject of our study and the experimental activity.

During the theoretical explanation, the various mathematical deductions found in Chapter 1 of this work were omitted, since they would require too much time, compromising the already short time available for such an activity, and also because a large proportion of the audience did not have the basic mathematical prerequisites to fully understand these passages. The experiment was actually carried out in the second week, i.e. the last two lessons set aside for dealing with the topic. However, before carrying out the experiment, a motivating text related to the subject was briefly discussed, as shown in ANNEX A. After reading and explaining this text, it was observed that several students had their interest aroused because of the experimental practice that would be carried out and the relationship between it and the motivating text. It should be noted that both the theoretical and practical lessons, with the experiment, were carried out in the classroom itself, since, although there was a science teaching laboratory, it was deactivated at the time of the experiment.

After reading and briefly discussing the motivating text mentioned above, the experimental activity began. Only one kit was used to carry out the experiment, which can be seen in Figure 5 of this chapter. We believe that having a larger number of kits could make the activity more interesting, productive and with greater didactic gains, since the class could be organized into groups, thus encouraging cooperative work among those involved. As an activity to be developed after the first experimental practice, each student could be asked to build their own kit, since, as already explained, many of the materials can be found in the student's home and are simple and inexpensive to make. In the case of this study, such an activity was not requested due to the shortage of time.

Due to the limited time available, the experiment was carried out for the first time by the teacher, so that the students could have a first contact with it. After this first run by the teacher, the students were invited to participate and interact more actively with the experimental apparatus and those who showed interest carried out the experiment already described for the other members of the class. In general, there was a greater interest and interaction among the students in the experimental activity compared to the traditional theoretical lessons. After the students had reproduced the experiment several times, a new discussion began on the conceptual aspects related to hydrostatics and Archimedes' principle.

The discussion began with a diagram on the board, similar to the one in Figure 1. At this point, the influence of the column of fluids (air and water) on the pressure exerted on the submarine or other submerged object was once again discussed (as this had already been done in the first two lessons). At this point there was no difficulty on the part of the students in understanding that the pressure is lower in the upper part of the submerged object than in the lower part.

Another aspect developed was the influence that the weight force exerts on the submerged object and the thrust, the latter also being a force, so that its action occurs from the bottom up, "pushing" the submarine towards the surface as opposed to the weight force, which acts by "pulling" the submerged object downwards. In this case, it was observed that the students had some difficulty in abstracting this situation. The students tend to relate this "new" force, which acts from the bottom upwards, to the normal force, perhaps because it is more familiar to them.

The students also discussed the fluctuation conditions of a body immersed in a fluid medium, which are described in more detail in Chapter 1 of this work. The discussion about the conditions of buoyancy is the moment that apparently raises the most questions and where the debate becomes the most interesting and productive. When the students were asked why the submarine rises when air is inflated into the balloon, the answer they gave was almost unanimous:

Air makes the submarine "lighter", professor!

Taking this answer as a starting point, the influence that the density of a fluid has on the aforementioned buoyancy conditions was highlighted. In this episode, it was discussed that, as soon as air is inflated into the balloon and, consequently, into the submarine, it has a lower average density than water, since the inflated air "expels" the water initially contained inside the submarine, taking its place, thus making it less dense than the liquid in which it is immersed. The consequence is that the thrust will be greater than the weight force, as already explained, which causes the submarine to return to the surface.

Another similar discussion, but just as interesting as the first, took place when the students were asked to provide possible solutions for causing the submarine to submerge completely, in other words, for them to propose a way of making the submarine sink completely to the bottom of the container. Once again, the answer was almost always unanimous:

You have to put "heavy" objects inside the submarine, Professor!

When asked what objects could be used for this purpose, the answers were generally:

Pieces of metal.

Small stones.

In this case, it is argued that, in such a situation, the average density of the submarine becomes much greater than that of the water, resulting in the weight force, directed downwards, being much greater than the thrust, causing the submarine to sink to the bottom of the container.

These were the discussions that took place in the classroom, although there are some other situations that could be worked on, but which were omitted in this research due to the short time available for such work. One possible example would be to take a quantitative approach, asking the students to make approximate measurements of the pressure values in the container, as a result of the height of the fluid column, as suggested above.

An important observation that was made during the experimental work concerns the differences in the understanding of the physical concepts covered, according to the different grades surveyed. In general, in the first years of the study, it was clear that it was difficult to abstract the physical phenomena discussed, and that they also had great difficulty in dealing with mathematical aspects, which can be explained in part by their immaturity in dealing with scientific aspects and also by the lack of prerequisites, a relatively common situation. In the second years, there is a slight improvement in the abstraction of physical concepts and in their mathematical treatment. Finally, in the third years, a considerable qualitative leap was observed, both in the abstraction of physical concepts and in mathematical treatment. These observations were made by applying a short questionnaire to gauge their grasp of the concepts covered, which is included in APPENDIX C of this work.

Chapter 3

Research Methodology

In order to carry out this work, a bibliographical and exploratory study was essentially used, since this seems to be more appropriate and aligned with the objectives of this study. This research was also supported and founded on the pillars of a qualitative approach, with the application of a research questionnaire, in order to bring to light a better understanding of the various problems exposed throughout the work and to verify some questions, related to the use of experiments as a didactic strategy, especially in the teaching of Physics, and the omission of certain content in high school classrooms.

With regard to the terminology "research", Manzato and Santos (2012, p. 3) teach us that:

> Research is an activity aimed at solving theoretical or practical problems using scientific processes. However, research is not the only way of obtaining knowledge and discoveries. Other means of accessing knowledge do not require the use of scientific processes, and although they are valid, they cannot be classified as research tasks. One of these means, which is highly recommended, is bibliographic consultation, which is characterized by clearing up minor doubts using documents.

In turn, Minayo (*apud* LIMA and MIOTO, 2007, p. 38), addressing this issue, points out that:

> Research is understood as a process in which the researcher has "an attitude and a theoretical practice of constant search that defines an intrinsically unfinished and permanent process", since he carries out an activity of successive approaches to reality, which has "a historical charge" and reflects positions in relation to reality.

From the point of view of its nature, this work can be classified as applied research, because according to Silva and Menezes (2005, p. 20), applied research "aims to generate knowledge for practical application and directed at solving specific problems. It involves local truths and interests". In this way, it fits in very well with what our work proposes, insofar as we seek to bring the philosophy of the low-cost physics teaching laboratory closer to the school community, facilitating the carrying out of simple experiments that are economically viable because they are cheap and easy to make, mainly aimed at school environments where there are no didactic teaching laboratories or, when present, they are deactivated because they are in poor working condition. We also wanted to encourage debate in the school environment and with the community about the importance of experimental practice as a didactic strategy necessary for the student's full education.

With regard to the approach adopted in this research, it can be classified as qualitative, as mentioned above. According to the concept taught by Malhotra (apud CHAER, DINIZ e RIBEIRO, 2011, p. 257): it is an "unstructured and exploratory research methodology, based on small samples that provide insights and understanding of the context of the problem". Silva and Menezes (2005) also teach us about qualitative research:

> [...] considers that there is a dynamic relationship between the real world and the subject, that is, an inseparable link between the objective world and the subjectivity of the subject that cannot be translated into numbers. The interpretation of phenomena and the attribution of meanings are basic to the qualitative research process. It does not require the use of statistical methods and techniques. The natural environment is the direct source for data collection and the researcher is the key instrument. It is descriptive. Researchers tend to analyze their data inductively. The process and its meaning are the main focus of the approach (SILVA e MENEZES, 2005, p. 20).

From the point of view of objectives, this is an exploratory study, because as Selltiz et. al. (*apud* Gil, 2002, p. 42) say about the subject:

> The aim of this type of research is to provide greater familiarity with the problem, with a view to making it more explicit or to constructing hypotheses. It can be said that the main purpose of this research is to improve ideas or discover intuitions. Its planning is therefore quite flexible, so that it allows the most varied aspects relating to the fact being studied to be considered. In most cases, this research involves: (a) a bibliographical survey; (b) interviews with people who have had practical experience with the problem being researched; and (c) analysis of examples that "stimulate understanding".

As far as the technical procedures are concerned, this is a bibliographical study, as mentioned above, because as Gil (2002, p. 44) wisely tells us:

> Bibliographical research is carried out on the basis of material that has already been prepared, consisting mainly of books and scientific articles. Although almost all studies require some kind of work of this nature, there are studies developed exclusively from bibliographic sources. A large part of exploratory studies can be defined as bibliographical research. Research into ideologies, as well as research that aims to analyze the various positions on a problem, is also usually developed almost exclusively through bibliographic sources.

Still on the subject of this study, it was carried out in a school belonging to the public education network of the Federal District in 2016. The questionnaire guiding the research - shown in APPENDIX D - was applied on September 8 and 9, 2016, in a school belonging to the Ceilândia Regional Education Coordination.

The choice of Ceilândia - the Administrative Region of the Federal District - is due to the fact that this satellite town is located in a generally low-income region. Its main characteristic is that it has a community made up mostly of individuals with very limited purchasing power. The precarious conditions of the public schools are also notable, with a few exceptions. All these characteristics fit in very well with the purpose of this study.

The people taking part in the survey are students duly enrolled in a public educational institution belonging to the aforementioned Administrative Region, in the 1st, 2nd and 3rd year of regular secondary school. The survey was completed by a total of 114 individuals during the period in question, making it a relatively small sample of the population. However, although the number of people surveyed was small, this is not a problem since the survey will basically be conducted using a qualitative approach.

In the specific case of the school where the research was carried out, it was observed that there is a space set aside for a science laboratory, although this space was deactivated at the time of the research, since there were no functional experimental apparatuses for carrying out experiments.

Even when faced with the situation described here, which is found in most of the country's public schools, we believe that experimental activity should not be omitted or neglected in any way at this stage of education, since it can contribute positively to a better education for the student. This is already known and is supported by the argument of Santos, Piassi and Ferreira (2004, p. 1):

Despite the importance of experimental activities in science education, most of the time science is still presented only through formulas, definitions and standardized exercises.

This approach removes important elements from the teaching of physics, such as the formation of various skills and habits, makes it difficult to learn concepts and, from an emotional point of view, alienates students from their interest in scientific knowledge. Experimentation, especially when carried out with simple materials that the student is able to manipulate and control, facilitates the learning of concepts, arouses interest and prompts an inquisitive attitude on the part of the student.

To collect the data, we used the questionnaire in APPENDIX D, which has 12 (twelve) questions. Of these, 5 (five) are open questions, which allowed respondents to express their opinion and position on certain questions, and 7 (seven) are closed questions, making it a mixed questionnaire, since respondents answered both open and closed questions. We would like to point out that a numerical scale was added to the 5 (five) open questions, from 0 (zero) to 10 (ten), in order to measure aspects and attitudes pertinent to our study. It should be noted that during and after the questionnaire was administered, confidentiality was fully guaranteed and the identity of those surveyed was preserved.

Chapter 4

Presentation and Analysis of Results

This chapter will present and briefly analyze the data obtained by applying the questionnaire in APPENDIX D. This data will be presented in two sets: the first relates to the profile of the public surveyed and the second relates to the didactic and pedagogical aspects.

4.1 Profile of the public surveyed

Initially, an analysis will be made of the profile of the people taking part in the research. At this point, the grade to which the students belong, the age group and the gender of the individuals surveyed will be analyzed.

4.1.1 Profile of the participating students in terms of grade.

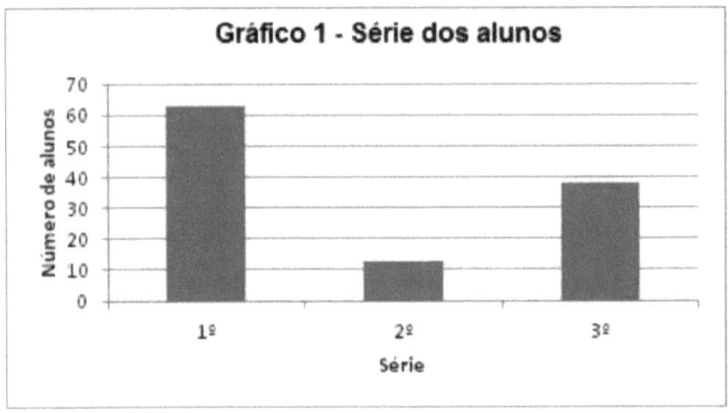

Figure 6 - Grade to which the students belong Source: author's own data

The graph shown in Figure 6 shows the number of students who were surveyed, according to the grade in which they are enrolled. We surveyed 63 students in the 1st year of secondary school, 13 students in the 2nd year of secondary school and 38 students in the 3rd year of this same stage of education, thus totaling a survey of 114 individuals, all duly enrolled in a public educational institution in the Federal District, which belongs to the Ceilândia Administrative Region.

4.1.2 Profile of the students in terms of age distribution.

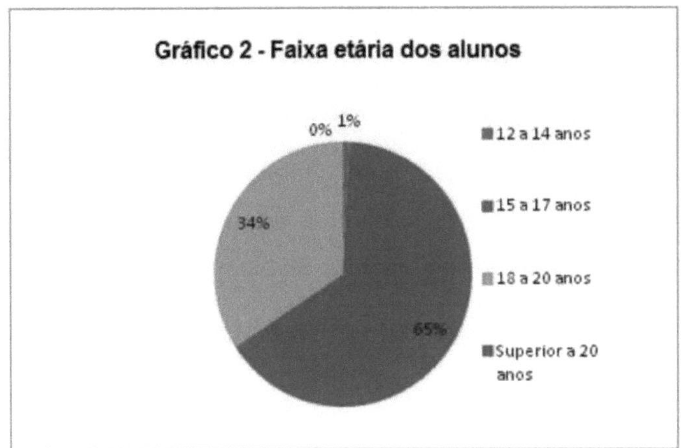

Figure 7 - Age range of the students Source: author's own work

The graph in Figure 7 shows that only 1% of the students surveyed were aged between 12 and 14. It also shows that 65% of the students belong to the 15-17 age group, i.e. the vast majority of the public surveyed. It also shows that 34% belong to the 18-20 age group. None of the students said they were over 20 years old. With this data, it can be seen that the vast majority are within acceptable educational parameters as far as the age-grade ratio is concerned.

4.1.3 Gender profile of the students.

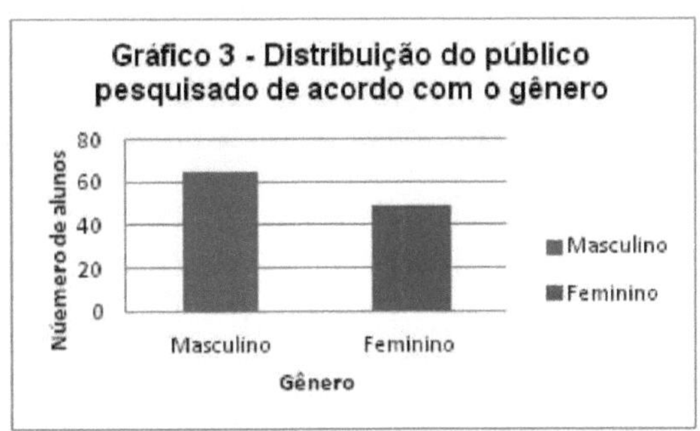

Figure 8 - Distribution of the public surveyed according to gender Source: author himself.

Figure 8 shows that the majority of the sample is made up of males, where 65 individuals were identified, and the remainder of females, i.e. 49 individuals, giving a total of 114 individuals.

4.2 Pedagogical didactic aspects related to the teaching of Physics and the use of the didactic laboratory.

This section will analyze data relating to more general aspects of physics teaching and also more specific aspects relating to the use of the teaching laboratory as a possible tool for improving teaching and learning processes.

4.2.1 Graph 4 - Have you had experimental physics lessons at any time during your school life?

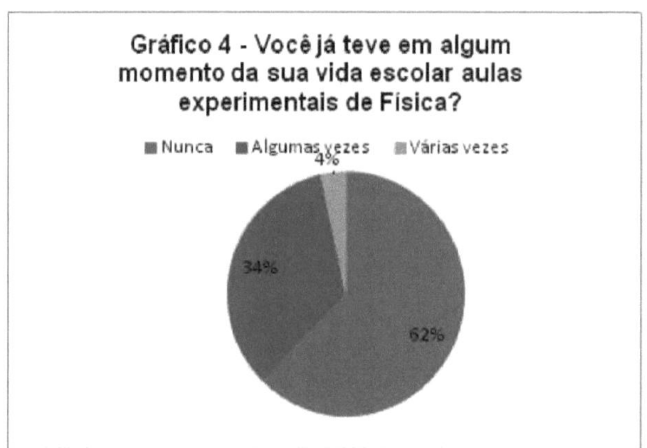

Figure 9 - Have you had experimental Physics lessons at any time during your school life?

Source: author himself.

The graph in Figure 9 shows that the majority of the students surveyed, i.e. 62% of the sample, claim never to have had an experimental Physics lesson. 32% of the public surveyed claim to have had experimental Physics lessons a few times during their academic life and only 4% of the sample claim to have had experimental Physics lessons several times during their school life.

This analysis confirms the existence of a problem that affects the teaching of Physics, namely the limited use of experimental activities in schools. One possible explanation for the almost total absence of experimental Physics lessons at this very important stage of education could be the lack of public investment, among others that could be listed. In this sense, Ferreira (1993, p. 97) argues very well when he points out that:

> The question of using the laboratory as an auxiliary resource for teaching physics is closely linked to numerous factors that can limit an in-depth analysis of the subject. For example, the lack of clear objectives for education in general, and for Physics in particular, inadequate teacher training, the school itself lacking the conditions for experimental activity, flat salaries, etc.

Thus, the explanations for the almost total absence of experimental physics lessons in secondary schools are varied and certainly more serious in the context of public schools, where the shortages and difficulties are usually greater.

4.2.2 Graph 5 - Do you like studying Physics?

The question - "Do you like studying Physics?" was accompanied by a scale from zero (0) to ten (10), so that the respondent could state whether or not they agreed with the question and we could measure aspects of our study more adequately. Together with the scale, the student was asked to briefly state their justification for the grade.

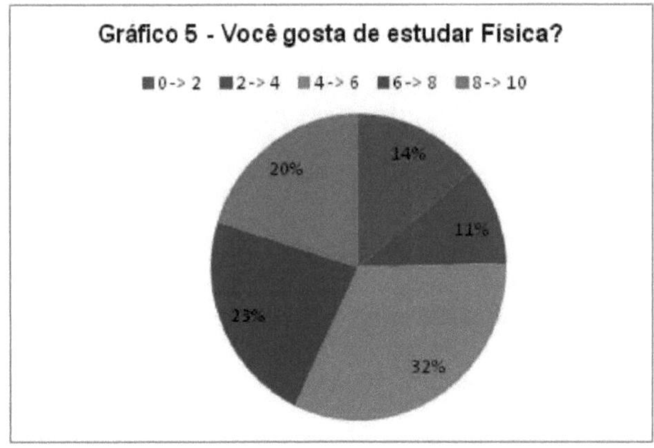

Figure 10 - Do you like studying Physics? Source: author.

The most significant results, along with a transcript of some interesting responses, are provided below:

- 16 students (14%) gave a score between 0 (zero) and 2

(two). In general, the most interesting answers are: I can't understand the subject.

I'm not good at exact sciences.

I don't think the subject contributed anything to my education.

- 12 of those surveyed (11%) gave a score between 2 (two)

and 4 (four). These responded in general as follows:

I don't like physics.

I don't like exact sciences.

I have trouble understanding the subject.

- 37 students (32%) gave the grade between 4 (four) and 6

(six). They answered the following in general:

There's a lot of theory and very little practice.

I don't like the article.

I can't understand the subject.

It has nothing to do with the profession I want to pursue.

It's a rather complicated subject.

I can't learn.

He has a lot of stones.

- 26 respondents (23%) gave a score between 6 (six) and

8 (eight). These students generally responded as follows:

I can relate my day-to-day life to the subject.

I'm interested in what Physics studies.

It's good to learn about physical phenomena.

It helps me in a future job.

The teacher explains well and makes the subject interesting.

I like calculations.

- 23 students (20%) gave the grade between 8 (eight) and 10 (ten). These students gave the following general answers:

It's an interesting story.

It's a way of acquiring knowledge.

I like calculations.

I like it because it deals with everyday phenomena and uses a lot of mathematics.

Physics has a very interesting content.

It's a subject that helps me a lot in life. I like the teacher's explanations.

From the information gathered and shown in the graph, it can be seen that there is a great deal of dissatisfaction and aversion in general to studying physics. This is for a variety of reasons, as the transcripts show. Only a small proportion of those surveyed seemed to be satisfied with the subject, attributing this to their sympathy with Mathematics, their numerical ability and the fact that they liked the lessons given by the teacher. One possible explanation for the dissatisfaction with this curricular component could be the difficulty that students at this stage of education have in associating the concepts learned in a theoretical way with their everyday world, in other words, it is difficult for them to establish relationships with the reality in which they are inserted and of which they are a part. Given this situation, the use of experiments, even low-cost ones like the one suggested in this paper, could make a considerable and positive contribution to improving the situation. As Santos, Piassi and Ferreira (2004, p. 6) wisely point out:

According to various authors, such as Lopes (2001) and Axt and Guimaraes (1985), it is natural to find among our high school students individuals who are still at the stage of concrete operations, defined by Piaget as one in which the individual is not yet capable of the abstractions necessary to understand the physics they are taught at school. Concepts such as force or energy require a degree of abstraction that our high school students have not yet reached.

Thus, experimental activities are a way of bringing these students closer to physics in a more concrete way. Even for students who have already reached the stage of formal operations, experimental activities, if not essential, are precisely and only at this stage that they present themselves as a possibility of complete exercise. They are therefore important for all students.

In this way, the authors confirm the benefits that experimental activities can bring to students in general.

4.2.3 Graph 6 - How motivated are you to study this subject?

This question was accompanied by a scale from zero (0) to ten (10), so that the respondent could state their level of motivation when studying Physics. In this context, zero (0) should be interpreted as "no motivation at all" and ten (10) as "totally motivated".

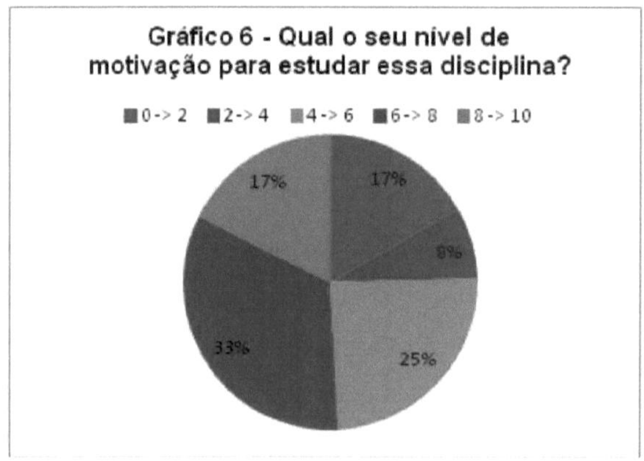

Figure 11 - How motivated are you to study this subject?

Source: author himself.

The data generated in the graph in Figure 11 seems to indicate that there is a certain degree of demotivation among the students about studying Physics. A percentage of 17% of those surveyed reported having no or little motivation to study this subject, with grades ranging from 0 (zero) to 2 (two). At the other extreme, 17% of the sample surveyed said they had a lot or a lot of motivation to study this subject, as they gave a score between 8 (eight) and 10 (ten) on the scale. 25% of those surveyed gave a score of between 4 (four) and 6 (six), i.e. they were "moderately" motivated and a further 33% of those surveyed gave a score of between 6 (six) and 8 (eight), i.e. they were "well" motivated to study this curricular component. With regard to this issue of lack of motivation on the part of the students, this fact can be explained partly as a result of the traditional teaching methodology, blackboard and chalk, which is widely adopted to the detriment of other teaching strategies where the student can adopt a more active stance in the construction of their knowledge. Clement, Custódio and Alves Filho (2015, p. 123) make a very good argument on this subject:

> We believe that the investigative aspect, peculiar to inquiry teaching, has the potential to arouse students' interest and greater engagement in the process of constructing their knowledge; culminating in greater motivational quality (autonomous motivation). Autonomous motivation is developed or improved to the extent that students realize their responsibility and role as learners. This aspect is often worked on in research-based teaching, as the aim is to set up problem situations that require the active participation of students in the development of solutions, which necessarily lead to the learning of new knowledge.

In this way, we understand that the use of experiments in Physics teaching, valuing investigative procedures in Physics learning, could contribute positively to the motivational aspect and to the acquisition of knowledge and development of various skills.

4.2.4 Graph 7 - Can you relate the concepts you learn in Physics in class to your everyday life?

This question was accompanied by a scale from zero (0) to ten (10), so that the respondent could say how much they can relate the content learned in the classroom to their daily lives. Zero (0) being the lowest level, i.e. the respondent can't relate any of the content to their daily lives, and ten (10) being the highest level, in which the respondent can relate all the content learned in the classroom to their daily lives.

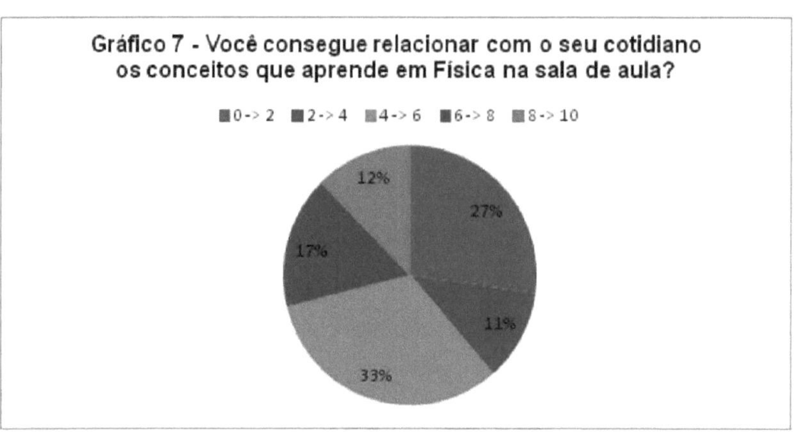

Figure 12 - Can you relate the concepts you learn in Physics in the classroom to your everyday life?

Source: author himself.

Analyzing the data shown in the graph in Figure 12, it can be seen that a significant percentage of the public surveyed (27%) gave a score between zero (0) and two (2), i.e. they were unable to relate the Physics content taught in the classroom to their daily lives, or very little. Only a small proportion, 12% of the questionnaires answered, said that they were able to associate the content learned in the classroom with their day-to-day lives. The majority of the sample studied - 33% - were able to relate in an "average" way the content taught in the classroom to their experience outside the classroom.

In the open-ended answers, the most frequently cited concepts were: displacement, speed, acceleration, movement, force, inertia, friction, collision between cars, gravity and temperature. It should be noted that a large proportion of those surveyed said they were unable to relate any of the content taught to their daily lives and another significant proportion, despite the instructions, preferred to leave the question unanswered. The fact that the respondents left their answers open-ended without any argument can be explained by the fact that they really do have some difficulty in relating physical concepts learned in the classroom to everyday life, a fact that can be explained by methodological issues, including the lack of practical and experimental activities.

4.2.5 Graph 8 - Can you relate the technologies of your daily life to the concepts you learn in Physics in class?

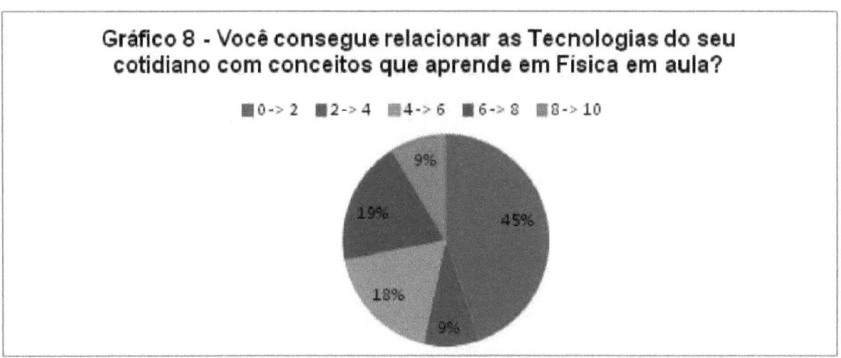

Figure 13 - Can you relate the technologies of your daily life to the concepts you learn in Physics in class?

Source: author himself.

Analyzing the data that generated the graph in Figure 13, it can be seen that a very significant percentage - 45% of the public surveyed - say that they can't relate everyday technologies to the subjects taught in the classroom at all, or very little. Only a small percentage - 8% - said they were able to relate the content taught in the classroom to everyday technologies well or fully. It should also be noted that, once again, the vast majority of individuals did not answer the open question, thus failing to give their opinion. This can be explained by the fact that the public surveyed finds it difficult to relate the technologies they use and experience in everyday life to Physics lessons.

4.2.6 Graph 9 - Do you believe that experimental Physics classes can make a positive contribution to your education?

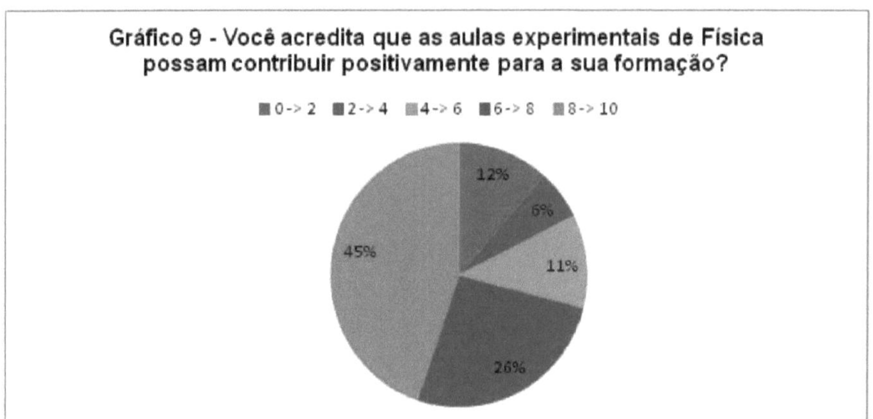

Figure 14 - Do you believe that experimental Physics classes can make a positive contribution to your education?

Source: author himself.

Looking at the graph in the figure, it is clear that the vast majority of the students surveyed - 45% of the individuals questioned - believe very strongly or completely that experimental Physics lessons can make a positive contribution in some way to their education. Another significant percentage - 26% of the sample surveyed - believe that experimental classes can make a positive contribution. Only 12% of the sample surveyed believed that experimental Physics lessons make little or no contribution to individuals' education.

The analysis of this data seems to suggest that the students somehow believe that experimental physics lessons really can make a beneficial contribution to their education.

4.2.7 Graph 10 - Do you believe that the use of physics experiments can facilitate understanding of the content covered in traditional physics classes?

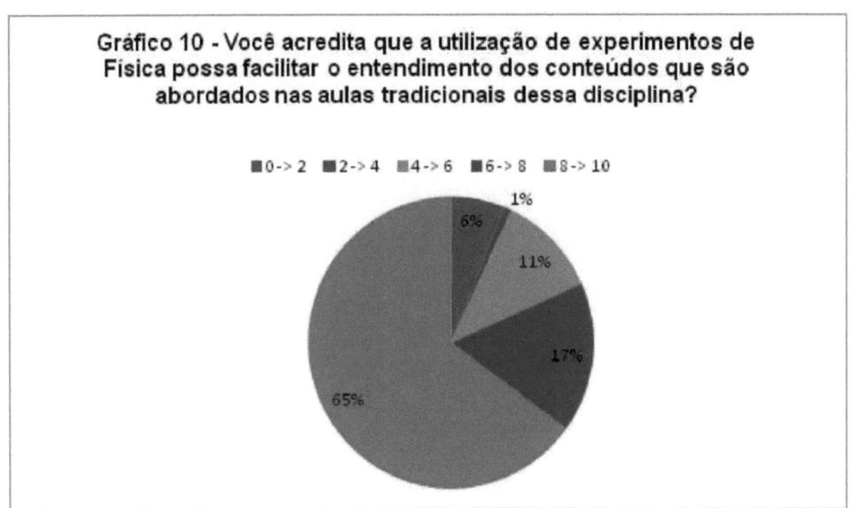

Figure 15 - Do you believe that using physics experiments can make it easier to understand the content covered in traditional physics classrooms?

Source: author himself.

Analysis of the data corresponding to the graph in Figure 15 shows that the vast majority of the students surveyed in the sample believe that in some way the use of experiments in Physics lessons is positive. They therefore believe that experimentation can contribute to their learning of Physics. Only a small proportion believe that the Physics laboratory does not contribute to their learning. Although unsolicited, some students expressed their opinion in writing in the questionnaire, with phrases such as:

Yes, the use of experiments helps me understand the subject better.

There should be more practical experiments and fewer lectures.

Experiments help me understand the subject better.

In this way, the questionnaire answers several questions that have been raised throughout this work, with regard to the presence of the didactic laboratory as an educational tool in the teaching-learning process.

4.2.8 Graph 11 - Was Hydrostatics taught at any time during your school life?

Figure 16 - Were you taught Hydrotactics at any time during your school life?

Source: author himself.

The results show that of the sample surveyed, only 12 (twelve) students said they had studied something related to Hydrostatics throughout their school life, i.e. only 11%. The vast majority, 102 students, said they had never had contact with hydrostatics, or 89% of those questioned.

It can be concluded that there seems to be a gap in the study of this subject in secondary schools, since the vast majority of the students surveyed said they had never had contact with hydrostatics. This may be due to various reasons. One of them is the limited number of hours allocated to Physics. Within this reality, the teacher has to choose certain content to the detriment of others, with a view to the quality of the content taught. Another possible explanation could be a misjudgment on the part of the teacher as to the relevance of the subject to the high school student's education.

4.2.9 Graph 12 - At some point in your school life, were you taught any content related to Archimedes' Principle?

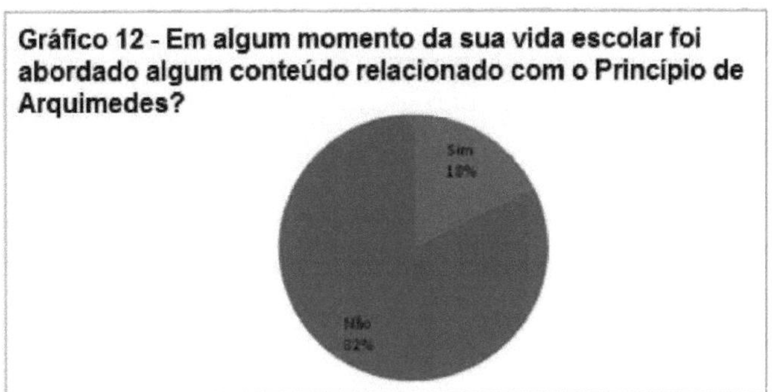

Figure 17 - At some point in your school life, were you taught any content related to Archimedes' Principle? Source: author.

Analyzing Graph 12, together with the information gathered through the questionnaire, it can be seen that the vast majority of the public surveyed - 82% - claim not to have had any contact with Archimedes' Principle. On the other hand, only 18% of those surveyed claim to have had some kind of contact with this subject at some point in their school life. Once again, the possible explanations for this may be due to the reduced workload allocated to Physics lessons, which means that teachers have to choose certain subjects over others.

Comparing the data analyzed in the graphs in Figures 11 and 12, we can see that there is a slight contrast. When respondents were asked whether they had studied hydrostatics, 89% said they had not had any contact with this subject and only 11% said they had studied it. When the respondents were asked if they had studied Archimedes' Principle, 82% said they had never had any contact with it, while 18% said they had had lessons on Archimedes' studies at some point in their school life.

Conclusions

In view of the discussions about the results obtained in this investigation, it can be concluded that the teaching laboratory for Physics (and Science in general) should be valued in secondary schools, especially the low-cost teaching laboratory, in view of its pedagogical importance for the student's education and the feasibility of its implementation even in school environments lacking instrumental resources for this practice, since, as already discussed, a large part of the country's public high schools lack this very important teaching resource.

To minimize this problem, we propose the implementation of a more practical and experimental approach, aimed at complementing the possible gaps left by traditional lectures, making it easier for students to act as the main actors in the construction of their knowledge and not just mere supporting actors, thus maximizing their cognitive development and raising the quality of the teaching and learning processes.

The suggested approach, i.e. the use of a low-cost experiment related to one of the topics of Hydrostatics, in the case of this work, Archimedes' Principle, was not arbitrary, since, as has already been extensively discussed throughout this research, there seems to be a large gap in this subject in secondary education, although it is an extremely important topic for the training of secondary school students.

We believe it goes without saying that the fact that Archimedes' Principle was chosen as an example to be implemented in a low-cost didactic laboratory does not limit the approach to this topic alone. There are countless possibilities for topics that can also be covered using this approach, including the possibility of covering topics from disciplines other than Physics. The example used in this paper illustrates just one possible situation from a whole range of possibilities. The hope is that, based on this example, a debate will be raised in the school community about the feasibility of using low-cost laboratories, and that this idea can be expanded to include other natural sciences and formal elements of mathematics.

The data collected for the discussion of the issues relevant to this work was done through the application of a survey questionnaire involving the participation of students belonging to the 1st, 2nd and 3rd years of high school, enrolled in a public school in the Federal District. The results reveal some possible answers to the questions raised throughout this study. The first data collected, regarding the age range of the public surveyed, showed that there were no marked distortions in terms of age-grade.

The data also shows that when the students were asked whether or not they had had any experimental lessons during their school life, the majority of the sample surveyed said they had never had any contact with experiments during their academic life. This can be explained by a number of reasons, such as the lack of public investment in education, the poor training of teachers or even the reduced number of hours allocated to science in public schools, which could make it impossible to work in this direction.

Other interesting conclusions concern the satisfaction students have in studying Physics and their motivation to study this curricular component. Once again, the vast majority of students said that they didn't like studying Physics and that they weren't very motivated to do so. Possible explanations for this phenomenon could be the current teaching model adopted, centered on expository lessons full of mathematical formulas, exercises and standardized problems, which do not value a more active attitude on the part of the students when it comes to investigating physical phenomena. One possible intervention that could be adopted to alleviate this widespread dissatisfaction would be the adoption of more practical, experimental and investigative classes, where students take a more active stance in the construction of their knowledge, abandoning the posture of listener.

We also obtained some interesting answers when we investigated whether students were able to relate what they learn in theoretical Physics classes to their everyday lives and whether they were able to associate the technologies of their daily lives with the concepts learned in Physics classes. Once again, the majority of the sample surveyed showed difficulties in both aspects. This fact becomes even clearer when analyzing the open-ended answers provided by the sample, since many of those surveyed were unable to provide adequate answers in this regard, often refraining from answering the question. A possible explanation for this, once again, lies in the traditional teaching model adopted for Physics classes, which values a very large number of theoretical classes to the detriment of practical classes with some experimentalism.

Other results point to the validity of using physics experiments in learning this curricular component. When the students were asked about this aspect, the vast majority of the sample said that the use of practical, experimental and investigative classes, etc. could make a positive contribution to understanding the phenomena covered in theoretical and traditional classes.

With regard to the presence of Hydrostatics and, consequently, Archimedes' Principle, the questionnaire guiding our research once again confirms what we had discussed throughout the work. The vast majority of our sample, which included students from all three years of secondary school, said that they had had no contact with this subject throughout their school life, confirming the hypothesis we had raised about negligence and omission in the treatment of this subject. This omission can be explained, in part, by the fact that there is a reduced timetable for teaching Physics, forcing teachers to value some subjects to the detriment of others.

Finally, in view of our analysis, the results we found are very much in line with what is discussed in the literature about the benefits of using the Physics teaching laboratory as a complementary resource to theoretical classes. In this way, we strongly believe that the use of experimental Physics classes, regardless of whether they are held in modern or low-cost laboratories, is promising in contributing to the apprehension of theoretical knowledge, while also adding skills that are indispensable for students' education.

References

ARAÙJO, M. S. T.; ABIB, M. L. V. S. **Experimental Activities in Physics Teaching: Different Approaches, Different Purposes**. Revista Brasileira de Ensino de Fisica, v. 25, n. 2, p. 176-194, Sâo Paulo, 2003.

BARBOSA, V. C.; BREITSCHAFT, A. M. S. **An experimental apparatus for the study of Archimedes' principle**. Revista Brasileira de Ensino de Fisica, v. 28, n. 1, p. 115-122, Sâo Paulo, 2006.

BRAGA, M.; GUERRA, A.; REIS, J. C. **Breve história da ciência moderna: Das Enlightes ao sonho do doutor Frankenstein.** v. 3, 2. ed., Rio de Janeiro: Zahar, 2011.

CHAER, G.; DINIZ, R. R. P.; RIBEIRO, E. A. **The questionnaire technique in educational research**. Evidência, v. 7, n. 7, p. 251-266, Araxà, 2011.

CLEMENT, L.; CUSTÓDIO, J. F.; FILHO, J. P. **Potential of Teaching by Investigation to Promote Autonomous Motivation in Scientific Education**. ALEXANDRIA Revista de Educaçâo em Ciência e Tecnologia, v. 8, n.1, p. 101-129, Santa Catarina, 2015.

DOCA, H. R.; BISCUOLA, G. J.; BÔAS, N. V. **Fisica**. v. 1, 1. ed., Sâo Paulo: Saraiva, 2010.

FEYERABEND, P. **Against the method**. Translated by Cezar Augusto Mortari. 2. ed. Sâo Paulo: Editora UNESP, 2011.

GIL, A. C. **Como elaborar projetas de pesquisa**. 4 ed. Sâo Paulo: Atlas, 2002.

GRANDINI, N. A.; GRANDINI, C. R. **The objectives of the didactic laboratory in the view of the students of the Physics degree course at UNESP-Bauru**. Brazilian Journal of Physics Teaching. Sociedade Brasileira de Ensino de Fisica, v. 26, n. 3, p. 251-256, Sâo Paulo, 2004.

LABURÙ, C. E.; ARRUDA. S. M; NARDI, R. **Pluralismo Metodológico no**

Ensino de Ciências. Ciência & Educaçâo, v. 9, n. 2, p. 247-260, Sâo Paulo, 2003.

LIMA, T. C. S.; MIOTO, R. C. T. **Methodological procedures in the construction of scientific knowledge: bibliographical research**. Rev. Kâtal, v. 10, n. esp., p. 37-45, Florianópolis, 2007.

MANZATO, A. J.; SANTOS, A. B. **Questionnaire design in quantitative research**. Department of Computer Science - IBILCE - UNESP, São Paulo, 2012.

NUSSENZVEIG, H. M. **Curso de Fisica Bàsica,** v. 2, 4 ed., Sâo Paulo: Edgard Blücher, 2004.

SANTOS, E. I.; FERREIRA, N. C.; PIASSI, L. P. C. **Activities Low Cost Experiments as Strategy Strategy Building Autonomy of Physics Teachers: An Experience in Continuing** Education. In: IX ENCONTRO NACIONAL DE PESQUISA EM ENSINO DE FÌSICA, 2004, Sâo Paulo: USP, 2004.

SILVA, E. L.; MENEZES, E. M. **Metodologia da pesquisa e elaboração de dissertaçâo**. 4. ed. rev. atual. Florianópolis: UFSC, 2005.

TORRES, S. **Navy changes rescue strategy**. Folha de S. Paulo, Sào Paulo, Dec. 28, 2000. 2000. Available at: <http://acervo.folha.uol.com.br/fsp/2000/12/28/2/>. Accessed on: August 15, 2016.

USP/CECAE. **The university and school science learning**. USP/BID Project, Training Science Teachers 1990-1993. São Paulo: USP, 1993.

ANNEX A

MOTIVATING TEXT - Navy changes rescue strategy

(Original article, published by the Folha de S. Paulo newspaper, on

December 28, 2000)

NAUFRÁGIO *Ação para trazer submarino à superfície tenta evitar desastre ambiental*

Marinha muda estratégia de resgate

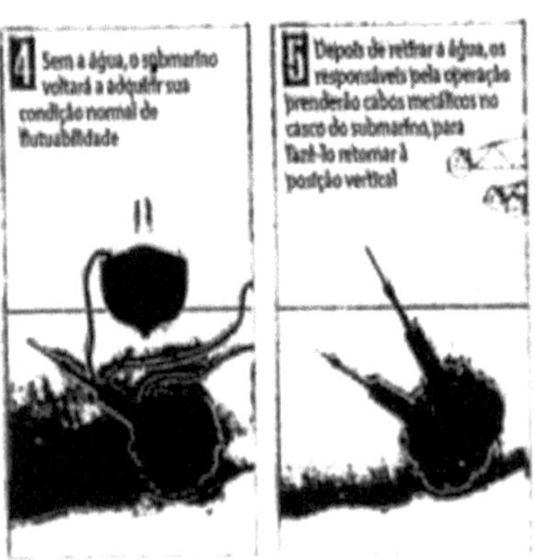

SERGIO TORRES
DA SUCURSAL DO RIO

Na tentativa de evitar um desastre ambiental, a Marinha decidiu mudar a estratégia de resgate do submarino Tonelero S-21, que naufragou no cais do Arsenal de Marinha, na baía de Guanabara, na noite de domingo.

A Marinha planejava trazer o submarino de volta à superfície injetando ar comprimido no interior do casco. Essa operação, se realizada, poderia espalhar óleo pela região da baía onde está o submarino — na ilha das Cobras, centro do Rio.

De acordo com nota divulgada ontem pela Marinha, existem dentro do submarino "borras de óleo" localizadas no porões, que estão inundados.

A nota é assinada pela capitão-de-corveta Aldner Peres de Oliveira, encarregada da Seção de Comunicação Social do 1º Distrito Naval (representação da Marinha no Estado do Rio). A Marinha informou ainda que existe a possibilidade de haver óleo lubrificante misturado à água.

O lubrificante teria sido derramado dos motores no momento em que o submarino se inclinou, ao bater em um declive no fundo do mar, a nove metros de profundidade. Não houve derramamento de óleo diesel, segundo a nota oficial. O diesel está lá nos tanques do Tonelero S-21.

Desde a noite do acidente, a Marinha cercou com bóias a área em volta do cais, a fim de impedir que o óleo eventualmente derramado se espalhasse pela baía.

O plano inicial da Marinha previa a injeção de ar comprimido dentro do submarino, para a expulsão da água por válvulas situadas na parte inferior do casco.

Os especialistas envolvidos no resgate do Tonelero S-21 desistiram do plano porque acham que haveria derramamento de óleo na baía. Um plano alternativo foi anunciado ontem. Ele consiste na remoção de uma escotilha no convés do submarino, com a fixação de uma espécie de duto acoplado a bombas de sucção.

As bombas deverão aspirar a água do interior do submarino, descarregando-a em uma barcaça coletora.

Previsto para terminar em 40 dias, o IPM (Inquérito Policial Militar) instaurado pela Marinha poderá ser prorrogado por mais 20 dias, caso os especialistas não cheguem a uma conclusão sobre as causas do afundamento do submarino.

A avaliação preliminar é que o acidente tenha sido causado por falhas mecânica e humana. Um defeito no sistema hidráulico da embarcação, que controla a abertura e o fechamento das comportas, não teria sido detectado a tempo pela tripulação.

A Marinha não está divulgando o teor dos depoimentos dos nove tripulantes (um oficial e oito praças) que estavam no Tonelero S-21 no momento do naufrágio. Os nomes não foram revelados. Os nove tripulantes não foram afastados do serviço diário.

Cerca de cem profissionais trabalham na operação de resgate, metade deles no local. A Marinha estima que o submarino voltará à superfície no início de janeiro.

APPENDIX A

MATERIAL USED TO BUILD THE SUBMARINE

To set up the experiment, we suggest a list of possible materials to be used. Of course, as you would expect from a low-cost physics laboratory, these materials can be easily adapted, depending on availability and the conditions in which the experiment will be carried out. A positive feature of this model of teaching laboratory is precisely this: the adaptability of materials and experiments according to the physical conditions of the environment in which the experiment will be carried out and the viability of materials.

Materials:

- 1 x 600 ml PET bottle (other bottles can be used)

sizes, including larger volumes. The choice varies according to the size of the container in which the experimental activity will take place).

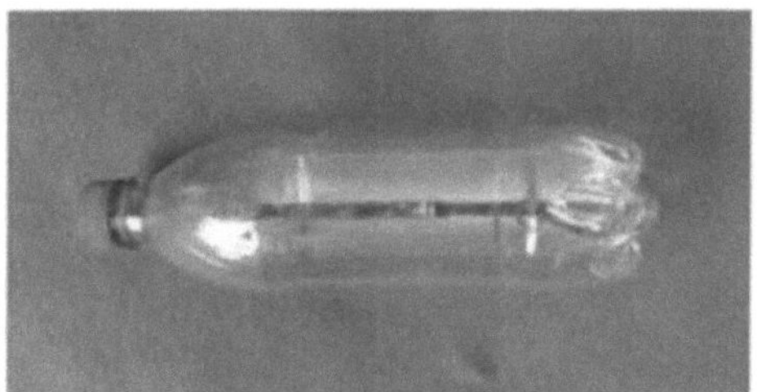

Figure 18 - Material used - 600 mi PET bottle Source: author's own work

- 1 balloon used for parties.

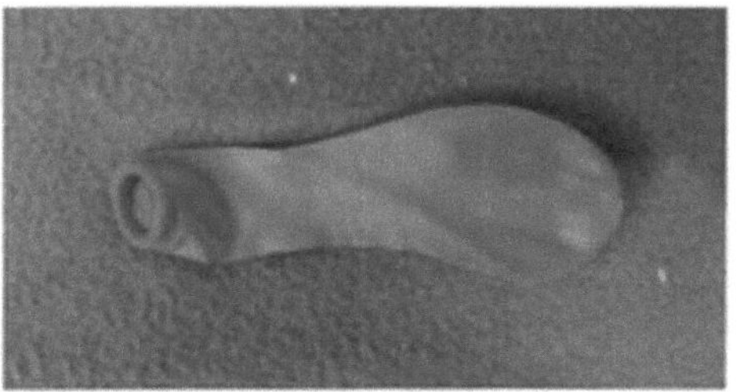

Figure 19 - Material used - Balloon Source: author's own.

- Adhesive tape, string, rubber bands or even small

plastic wires (the choice is at the discretion of the applicator). experience and availability of material)

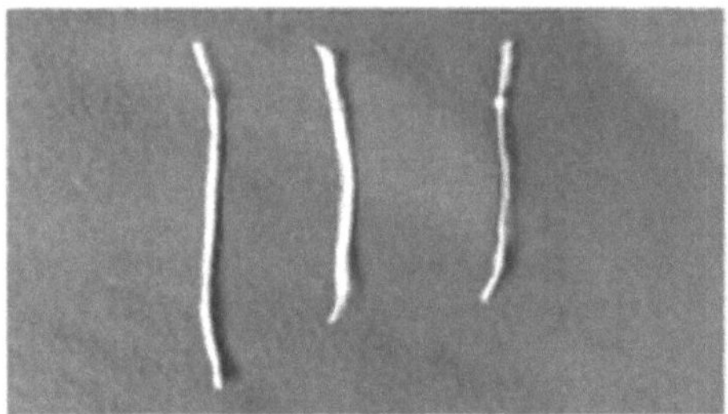

Figure 20 - Material used - Plastic wires Source: author's own work.

- Rubber hose or tourniquet 60 cm long or longer.

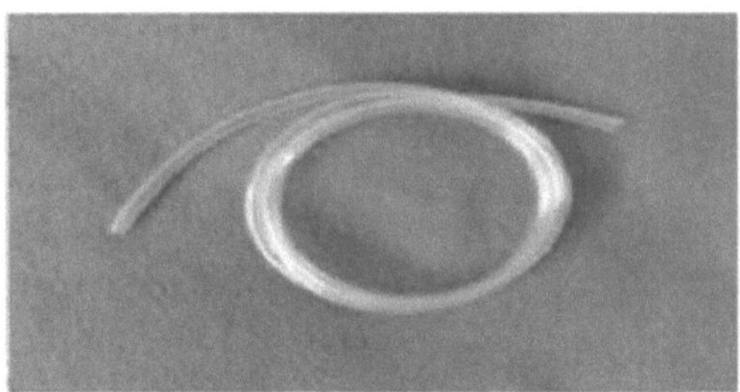

Figure 21 - Material used - Rubber hose Source: author's own.

- Nails, scissors and similar tools.

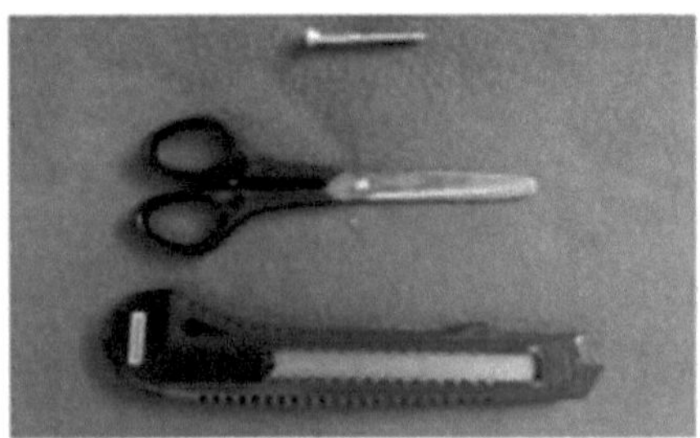

Figure 22 - Material used - Nails, scissors and similar tools.

Source: author.

- A transparent container for a considerable volume of water. You can use a small aquarium (a more expensive solution), basins, tanks, container boxes or something similar. Once again, the choice of container varies according to the conditions and the viability of the material.

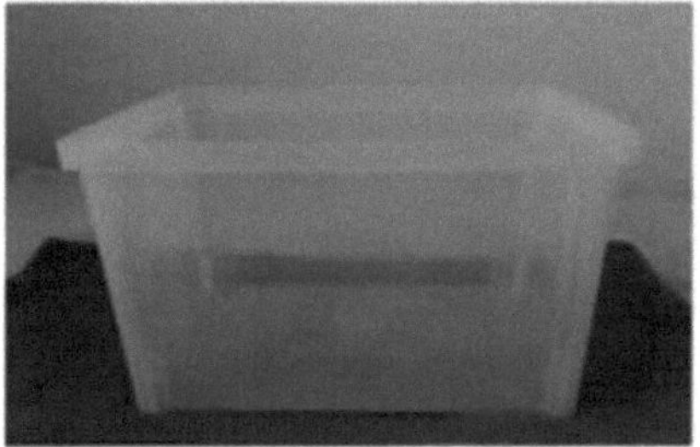

Figure 23 - Material used - Transparent container.

Source: author.

APPENDIX B

SCRIPT FOR SETTING UP THE EXPERIMENT

The details involved in the construction of the experimental apparatus, the subject of this study, are presented here.

Procedures:

1. Several holes are drilled in the PET bottle, which in the experiment has the function of allowing liquid to enter and leave the submarine. Next, a single hole is drilled in the PET bottle cap, where the hose with the balloon will be inserted. The various holes made in the experimental apparatus can be obtained using a nail, screw or equivalent instrument.

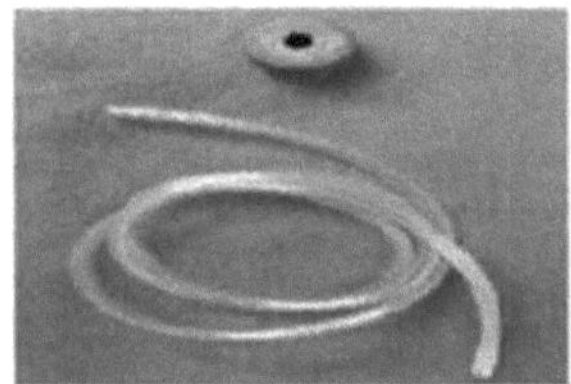

Figure 24 - Script for setting up the experiment Source: author's own work.

2. Using adhesive tape, string, rubber bands or small plastic wires, attach the balloon to one end of the hose so that no air escapes, or at least minimizes it.

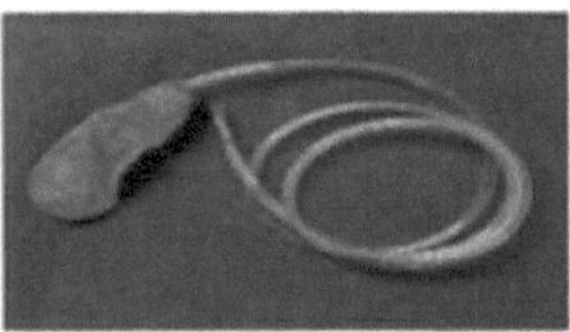
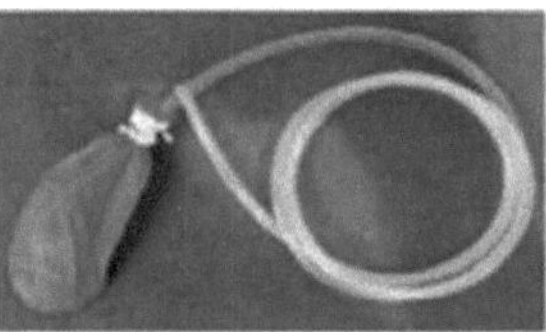

Figure 25 - Script for setting up the experiment.

Source: author.

3. Insert the balloon, which should already be attached to one end of the hose, into the PET bottle.

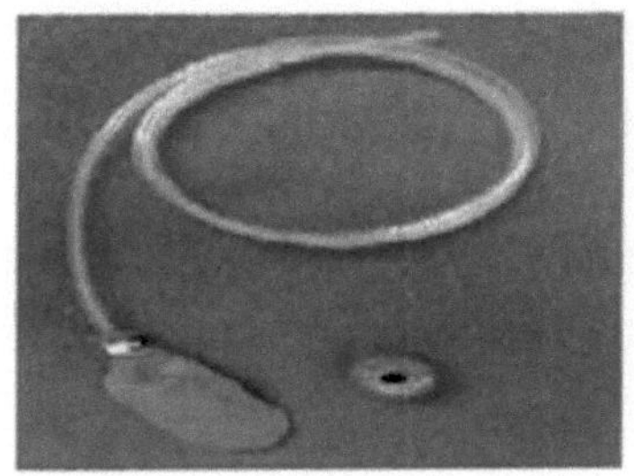
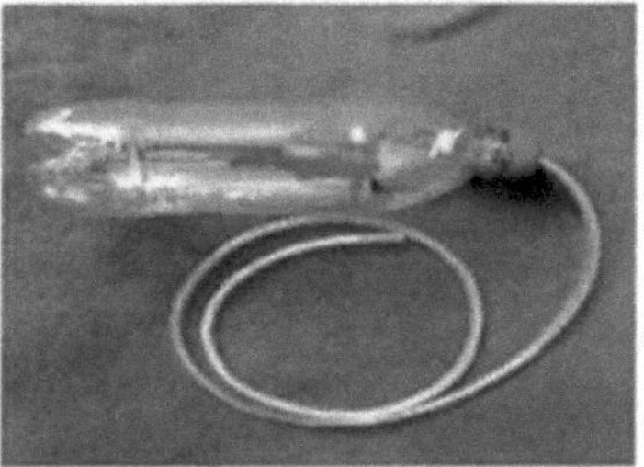

Figure 26 - Script for setting up the experiment.

Source: author.

4. Fill the container to be used with water.

APPENDIX C

ROUTE AND QUESTIONS FOR THE DEVELOPMENT OF THE EXPERIMENTAL ACTIVITY

1. Identify which physical concepts are involved in Archimedes' Principle and which would be involved in the proposed experiment.
2. Why does the submarine rise when the balloon is inflated?
3. What force is responsible for the submarine rising to the surface?
4. If you fill the balloon gradually, you can see that the submarine's speed of ascent increases as the volume of the balloon increases. What can you conclude from this observation?
5. Submarines can navigate on the surface of the sea or submerge and move vertically within it. Research and answer how this process works.
6. With regard to how submarines work, would it be correct to say that the process of rising and falling is due to an increase or decrease in density? Justify your answer.

APPENDIX D

QUESTIONNAIRE

1) What grade of secondary school are you currently attending?

( )1st Year

( )2nd Year

( )3rd Year

2) How old are you ?

( )12 to 14 years old

( )15 to 17years

( )18 to 20 years

( )More than 20 years

3) What's your gender?

( ) Male

( )Female

4) Have you ever had experimental physics lessons at school?

( )Never

( )Sometimes

( )Several times

Using a scale from zero (0) to ten (10), answer the following questions:

5) Do you like studying physics?

R:

Why?

6) How motivated are you to study this subject?

R:

7) Can you relate the concepts you learn in Physics in the classroom to your everyday life?

R:

Name a relationship: ____

8) Can you relate the technologies of your daily life to the concepts you learn in Physics in class?

R:

Name a relationship: ____

9) Do you believe that experimental physics lessons can make a positive contribution to your education?

R:

Justify your answer:

10) Do you believe that using physics experiments can make it easier to understand the content covered in traditional physics classes?

R:

11) Were you taught Hydrostatics at any point in your school life?

( ) Yes

( ) No

12) At any point in your school life have you been taught anything related to Archimedes' Principle?

( ) Yes

( ) No

Printed by Books on Demand GmbH, Norderstedt / Germany